酒店大堂设计
Hotel Lobby Design

DAM工作室 主编

华中科技大学出版社
http://www.hustp.com
中国·武汉

目录

中文名称	页码	英文名称
圣塞瓦斯蒂安玛丽亚克里斯蒂娜酒店	006	Hotel Maria Cristina, a Luxury Collection Hotel, San Sebastian
阿布扎比萨迪亚特岛瑞吉度假村	010	The St. Regis Saadiyat Island Resort, Abu Dhabi
阿布扎比首都希尔顿大酒店	011	Hilton Capital Grand Abu Dhabi
阿尔科巴尔艾美酒店	012	Le Méridien Al Khobar
阿格拉莫卧儿ITC酒店	018	ITC Mughal Agra
艾斯瓦兰斯苏拉特港威酒店	024	The Gateway Hotel Athwalines Surat
澳门金沙城中心康莱德酒店	025	Conrad Macao, Cotai Central
澳门喜来登金沙城中心酒店	026	Sheraton Macao Hotel, Cotai Central
澳门悦榕庄	030	Banyan Tree Macau
巴尔港瑞吉度假酒店	034	The St. Regis Bal Harbour Resort
巴库费尔蒙酒店	035	Fairmont Baku
巴厘岛金巴兰艾美酒店	036	Le Méridien Bali Jimbaran
巴厘岛努沙杜拉古娜豪华精选度假酒店	038	The Laguna, a Luxury Collection Resort & Spa, Nusa Dua, Bali
巴厘岛瑞吉度假酒店	040	The St. Regis Bali Resort
巴厘岛水明漾W度假酒店	042	W Retreat & Spa Bali - Seminyak
（上图）巴统喜来登酒店	043	(Above) Sheraton Batumi Hotel
（下图）布尔萨喜来登酒店	043	(Below) Sheraton Bursa Hotel
巴黎班克酒店	044	Banke Hotel, Paris
巴亚以塔港希尔顿度假酒店	046	Hilton Puerto Vallarta Resort
（上图）长白山假日酒店	047	(Above) Holiday Inn Changbaishan
（下图）长白山万达威斯汀度假酒店	047	(Below) The Westin Changbaishan Resort
芭堤雅希尔顿酒店	048	Hilton Pattaya
班加罗尔ITC皇家花园豪华精选酒店	050	ITC Gardenia, a Luxury Collection Hotel, Bengaluru
北京嘉里大酒店	052	Kerry Hotel, Beijing
贝尔格莱德大都会皇宫饭店	054	Metropol Palace, Belgrade
槟城香格里拉沙洋度假酒店	056	Shangri-La's Rasa Sayang Resort & Spa, Penang
长白山万达喜来登度假酒店	057	Sheraton Changbaishan Resort
长春静月潭益田喜来登酒店	058	Sheraton Changchun Jingyuetan Hotel
常州武进九洲喜来登酒店	060	Sheraton Changzhou Wujin Hotel
大阪瑞吉酒店	064	The St. Regis Osaka
迪拜阿联酋购物中心喜来登酒店	066	Sheraton Dubai Mall of the Emirates Hotel
迪拜棕榈岛华尔道夫酒店	067	Waldorf Astoria Dubai - Palm Jumeirah
迪拜马奎斯JW万豪酒店	068	JW Marriott Marquis Dubai
迪拜卓美亚帆船酒店	072	Burj Al Arab
迪拜棕榈岛One&Only度假村	078	One&Only The Palm
迪拜棕榈岛卓美亚斯布尔宫酒店	084	Jumeirah Zabeel Saray
东方红树林安纳塔拉水疗酒店	090	Eastern Mangroves Hotel & Spa
多哈W酒店及公寓	094	W Doha Hotel & Residences
多哈君悦酒店	098	Grand Hyatt Doha

CONTENTS

中文	页码	英文
阿布歇隆区巴库 JW 万豪酒店	099	JW Marriott Absheron Baku
多哈瑞吉酒店	100	The St. Regis Doha
法拉克努马宫泰姬酒店	102	Taj Falaknuma Palace, Hyderabad
佛罗伦萨利马斯克雷别墅酒店	108	Villa Le Maschere Resort, Florence
佛山希尔顿酒店	110	Hilton Foshan
福州万达威斯汀酒店	111	The Westin Fuzhou Minjiang
哥印拜陀艾美酒店	112	Le Méridien Coimbatore
广州 W 酒店	116	W Guangzhou
古尔默尔格开柏喜马拉雅度假村及水疗中心	120	The Khyber Himalayan Resort & Spa, Gulmarg
广州花都合景喜来登度假酒店	122	Sheraton Guangzhou Huadu Resort
海得拉巴威斯汀酒店	126	The Westin Hyderabad Mindspace
杭州西溪悦榕庄	127	Banyan Tree Hangzhou
海牙戴斯因德斯豪华精选酒店	128	Hotel Des Indes, a Luxury Collection Hotel, The Hague
杭州千岛湖滨江希尔顿度假酒店	130	Hilton Hangzhou Qiandao Lake Resort
杭州西溪喜来登度假酒店	136	Sheraton Hangzhou Wetland Park Resort
湖州喜来登温泉度假酒店	137	Sheraton Huzhou Hot Spring Resort
惠州金海湾喜来登度假酒店	138	Sheraton Huizhou Beach Resort
吉隆坡喜来登帝王酒店	139	Sheraton Imperial Kuala Lumpur Hotel
旧金山 W 酒店	142	W San Francisco
昆明七彩云南温德姆至尊豪廷大酒店	143	Wyndham Grand Plaza Royale Colorful Yunnan Kunming
旧金山瑞吉酒店	144	The St. Regis San Francisco
卡斯尔夏克华尔道夫酒店	146	Waldorf Astoria Jeddah – Qasr Al Sharq
拉萨瑞吉度假酒店	148	The St. Regis Lhasa Resort
兰珂悦椿度假村	150	Angsana Lang Co
鹿谷瑞吉酒店	152	The St. Regis Deer Valley
马来西亚柔佛盛贸饭店	158	Traders Hotel, Puteri Harbour, Johor, Malaysia
利马威斯汀酒店及会议中心	159	The Westin Lima Hotel & Convention Center
罗得岛喜来登度假酒店	160	Sheraton Rhodes Resort
（下图）伦敦希尔顿温布利酒店	161	(Below) Hilton London Wembley
马尔代夫瑞提拉岛 One&Only 度假村	162	One&Only Reethi Rah, Maldives
迈阿密南海滩 W 酒店	164	W South Beach
秘鲁唐波德尔英卡度假酒店	166	Tambo del Inka Resort & Spa, Valle Sagrado
莫里斯岛艾美酒店	169	Le Méridien Ile Maurice
明尼阿波利斯威斯汀酒店	170	The Westin Minneapolis
纳泰攀牙麦拷梦水疗度假酒店	172	Maikhao Dream Resort & Spa, Natai, Phang Nga
（上图）南京威斯汀大酒店	174	(Above) The Westin Nanjing
（下图）米兰马尔蓬萨喜来登酒店及会议中心	174	(Below) Sheraton Milan Malpensa Airport Hotel & Conference Center
纽约君悦酒店	175	Grand Hyatt New York
宁波威斯汀酒店	176	The Westin Ningbo

中文名称	页码	英文名称
纽约瑞吉酒店	178	The St. Regis New York
欧文兰德公园会议中心喜来登酒店	179	Sheraton Overland Park Hotel at the Convention Center
普林斯维尔瑞吉度假酒店	180	The St. Regis Princeville Resort
钦奈 ITC 大佐拉酒店	182	ITC Grand Chola, a Luxury Collection Hotel, Chennai
钦奈贝拉奇瑞威斯汀酒店	184	The Westin Chennai Velachery
秦皇岛北戴河华贸喜来登酒店	185	Sheraton Qinhuangdao Beidaihe Hotel
青岛胶州绿城喜来登酒店	186	Sheraton Qingdao Jiaozhou Hotel
（上图）青岛鲁商凯悦酒店	187	(Above) Hyatt Regency Qingdao
（下图）上海外高桥喜来登酒店	187	(Below) Sheraton Shanghai Waigaoqiao Hotel
新加坡皮克林宾乐雅酒店	188	Parkroyal on Pickering, Singapore
（上图）天津海河悦榕庄	189	(Above) Banyan Tree Tianjin Riverside
（下图）乌干沙悦榕庄	189	(Below) Banyan Tree Ungasan
清远狮子湖喜来登度假酒店	190	Sheraton Qingyuan Lion Lake Resort
曲阜香格里拉大酒店	191	Shangri-La Hotel, Qufu
塞维利亚阿方索十三世豪华精选酒店	194	Hotel Alfonso XIII, a Luxury Collection Hotel, Seville
三亚海棠湾凯宾斯基酒店	196	Kempinski Hotel Haitang Bay Sanya
阿尔布费拉阿尔加维喜来登豪华精选酒店	198	Sheraton Algarve, a Luxury Collection Hotel, Albufeira
三亚海棠湾喜来登假酒店	200	Sheraton Sanya Haitang Bay Resort
三亚亚龙湾瑞吉度假酒店	202	The St. Regis Sanya Yalong Bay Resort
三亚御海棠豪华精选度假酒店	204	The Royal Begonia, a Luxury Collection Resort, Sanya
沙特阿拉伯王国麦加钟塔皇家费尔蒙酒店	206	Makkah Clock Royal Tower, a Fairmont Hotel
上海滴水湖皇冠假日酒店	208	Crowne Plaza Shanghai Harbour City
上海卓美亚喜玛拉雅酒店	211	Jumeirah Himalayas Hotel Shanghai
深圳东部华侨城瀑布酒店	212	Otique Aqua Hotel Shenzhen
（上图）深圳君悦酒店	213	(Above) Grand Hyatt Shenzhen
（下图）石家庄希尔顿酒店	213	(Below) Hilton Shijiazhuang
深圳瑞吉酒店	214	The St. Regis Shenzhen
神州半岛喜来登度假酒店	215	Sheraton Shenzhou Peninsula Resort
阿布扎比盖斯尔阿萨拉沙漠度假村	216	Qasr Al Sarab Desert Resort by Anantara, Abu Dhabi
三亚海棠湾民生威斯汀度假酒店	217	The Westin Sanya Haitang Bay Resort
首尔 D 立方市喜来登酒店	218	Sheraton Seoul D Cube City Hotel
首尔华克山庄 W 酒店	219	W Seoul - Walkerhill
苏梅岛 W 酒店	220	W Retreat Koh Samui
索菲特曼谷特色酒店	222	Sofitel So Bangkok
台北 W 酒店	224	W Taipei
泰国清莱艾美度假酒店	226	Le Méridien Chiang Rai Resort, Thailand
天津瑞吉金融街酒店	228	The St. Regis Tianjin
万绿湖东方国际酒店	232	Oriental International Hotel, Wanlv Lake, Heyuan
威尼斯丹尼利豪华精选酒店	234	Hotel Danieli, a Luxury Collection Hotel, Venice

中文名称	页码	英文名称
韦尔比耶 W 酒店	238	W Verbier
维也纳布里斯托尔豪华精选酒店	242	Hotel Bristol, a Luxury Collection Hotel, Vienna
香港愉景湾酒店	244	Auberge Discovery Bay Hong Kong
（上图）西双版纳安纳塔拉度假酒店	245	(Above) Anantara Xishuangbanna Resort & Spa
（下图）武汉万达威斯汀酒店	245	(Below) The Westin Wuhan Wuchang
西安赛瑞喜来登大酒店（城北）	246	Sheraton Xi'an North City Hotel
西双版纳避寒皇冠假日度假酒店	248	Crowne Plaza Resort Xishuangbanna
仙台威斯汀酒店	255	The Westin Sendai
厦门威斯汀酒店	256	The Westin Xiamen
香格里拉 Barr Al Jissah 度假酒店	258	Shangri-La's Barr Al Jissah Resort & Spa, Sultanate of Oman
香格里拉莎利雅度假酒店	262	Shangri-La's Rasa Ria Resort, Kota Kinabalu
新德里 ITC 孔雀王朝豪华精选酒店	264	ITC Maurya, a Luxury Collection Hotel, New Delhi
新加坡圣淘沙湾 W 酒店	266	W Singapore - Sentosa Cove
新加坡万豪酒店	268	Singapore Marriott Hotel
新加坡威斯汀酒店	270	The Westin Singapore
休斯顿市中心威斯汀酒店	273	The Westin Houston Downtown
雅典乔治国王豪华精选酒店	280	King George, a Luxury Collection Hotel, Athens
雅加达苏丹皇宫广场酒店	286	Keraton at the Plaza, Jakarta
亚特兰大瑞吉酒店	288	The St. Regis Atlanta
亚特兰大市中心 W 酒店	292	W Atlanta - Downtown
伊斯坦布尔艾美酒店	294	Le Méridien Istanbul Etiler
宜兴万达艾美酒店	296	Le Méridien Yixing
印度斋普尔费尔蒙酒店	298	Fairmont Jaipur, India
镇江万达喜来登酒店	300	Sheraton Zhenjiang Hotel
芝加哥北溪喜来登酒店	302	Sheraton Chicago Northbrook Hotel
重庆北碚悦榕庄	304	Banyan Tree Chongqing Beibei
新加坡浮尔顿海湾酒店	307	The Fullerton Bay Hotel Singapore
卓美亚 Bilgah 海滩酒店	308	Jumeirah Bilgah Beach Hotel
卓美亚 Etihad Tower 酒店	312	Jumeirah at Etihad Towers
弗雷明汉喜来登会议中心酒店	314	Sheraton Framingham Hotel & Conference Center
马尔代夫满月岛喜来登度假酒店	315	Sheraton Maldives Full Moon Resort & SPA
迈阿密蒙德里安南部海滩酒店	316	Mondrian South Beach Hotel Miami
卡尔加里奥克莱尔喜来登套房酒店	317	Sheraton Suites Calgary Eau Claire
曼谷 W 酒店	318	W Bangkok
蒙特利尔 W 酒店	320	W Montréal
明尼阿波利斯钱伯斯艾美酒店	321	Le Méridien Chambers Minneapolis
柏林 Axo Light in Rome and Bonaldo	322	Axo Light in Rome and Bonaldo in Berlin
库斯科印加宫殿豪华精选酒店	324	Palacio del Inka, a Luxury Collection Hotel, Cusco

圣塞瓦斯蒂安玛丽亚克里斯蒂娜酒店
Hotel Maria Cristina, a Luxury Collection Hotel, San Sebastian

圣塞瓦斯蒂安玛丽亚克里斯蒂娜酒店
Hotel Maria Cristina, a Luxury Collection Hotel, San Sebastian

阿布扎比萨迪亚特岛瑞吉度假村 *The St. Regis Saadiyat Island Resort, Abu Dhabi*

Hilton Capital Grand Abu Dhabi 阿布扎比首都希尔顿大酒店

阿尔科巴尔艾美酒店 *Le Méridien Al Khobar*

阿尔科巴尔艾美酒店 *Le Méridien Al Khobar*

阿尔科巴尔艾美酒店 Le Méridien Al Khobar

阿格拉莫卧儿 ITC 酒店 *ITC Mughal Agra*

阿格拉莫卧儿 ITC 酒店
ITC Mughal Agra

阿格拉莫卧儿 ITC 酒店
ITC Mughal Agra

艾斯瓦兰斯苏拉特港威酒店 *The Gateway Hotel Athwalines Surat*

Conrad Macao, Cotai Central 澳门金沙城中心康莱德酒店

澳门喜来登金沙城中心酒店 *Sheraton Macao Hotel, Cotai Central*

澳門喜來登金沙城中心酒店
Sheraton Macao Hotel, Cotai Central

澳门悦榕庄
Banyan Tree Macau

澳门悦榕庄 *Banyan Tree Macau*

巴尔港瑞吉度假酒店 *The St. Regis Bal Harbour Resort*

Fairmont Baku 巴库费尔蒙酒店

巴厘岛金巴兰艾美酒店
Le Méridien Bali Jimbaran

巴厘岛努沙杜拉古娜豪华精选度假酒店 *The Laguna, a Luxury Collection Resort & Spa, Nusa Dua, Bali*

巴厘岛瑞吉度假酒店
The St. Regis Bali Resort

041

巴厘岛水明漾 W 度假酒店 *W Retreat & Spa Bali - Seminyak*

(Above) Sheraton Batumi Hotel （上图）巴统喜来登酒店　　*(Below) Sheraton Bursa Hotel* （下图）布尔萨喜来登酒店

巴黎班克酒店 *Banke Hotel, Paris*

巴亚以塔港希尔顿度假酒店 *Hilton Puerto Vallarta Resort*

(Above) Holiday Inn Changbaishan（上图）长白山假日酒店　　　*(Below) The Westin Changbaishan Resort*（下图）长白山万达威斯汀度假酒店

芭堤雅希尔顿酒店 *Hilton Pattaya*

班加罗尔 ITC 皇家花园豪华精选酒店 *ITC Gardenia, a Luxury Collection Hotel, Bengaluru*

051

北京嘉里大酒店 *Kerry Hotel, Beijing*

053

贝尔格莱德大都会皇宫饭店
Metropol Palace, Belgrade

槟城香格里拉沙洋度假酒店 *Shangri-La's Rasa Sayang Resort & Spa, Penang*

Sheraton Changbaishan Resort 长白山万达喜来登度假酒店

长春静月潭益田喜来登酒店 *Sheraton Changchun Jingyuetan Hotel*

常州武进九洲喜来登酒店 *Sheraton Changzhou Wujin Hotel*

常州武进九洲喜来登酒店
Sheraton Changzhou Wujin Hotel

大阪瑞吉酒店 *The St. Regis Osaka*

065

迪拜阿联酋购物中心喜来登酒店 *Sheraton Dubai Mall of the Emirates Hotel*

Waldorf Astoria Dubai - Palm Jumeirah 迪拜棕榈岛华尔道夫酒店

迪拜马奎斯 JW 万豪酒店 *JW Marriott Marquis Dubai*

迪拜马奎斯 JW 万豪酒店
JW Marriott Marquis Dubai

迪拜卓美亚帆船酒店
Burj Al Arab

迪拜卓美亚帆船酒店 *Burj Al Arab*

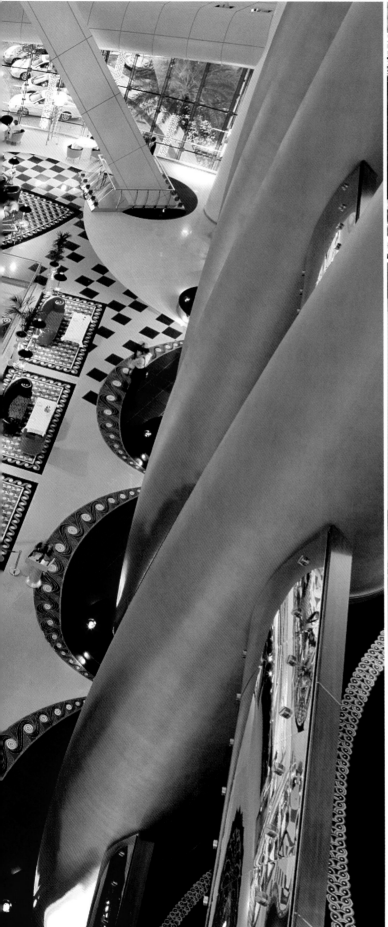

迪拜卓美亚帆船酒店
Burj Al Arab

迪拜棕榈岛 One&Only 度假村 *One&Only The Palm*

迪拜棕榈岛 One&Only 度假村
One&Only The Palm

迪拜棕榈岛One&Only度假村

迪拜棕榈岛卓美亚斯布尔宫酒店
Jumeirah Zabeel Saray

迪拜棕榈岛卓美亚斯布尔宫酒店 *Jumeirah Zabeel Saray*

迪拜棕榈岛卓美亚斯布尔宫酒店 *Jumeirah Zabeel Saray*

东方红树林安纳塔拉水疗酒店 *Eastern Mangroves Hotel & Spa*

东方红树林安纳塔拉水疗酒店 *Eastern Mangroves Hotel & Spa*

多哈 W 酒店及公寓 *W Doha Hotel & Residences*

多哈 W 酒店及公寓
W Doha Hotel & Residences

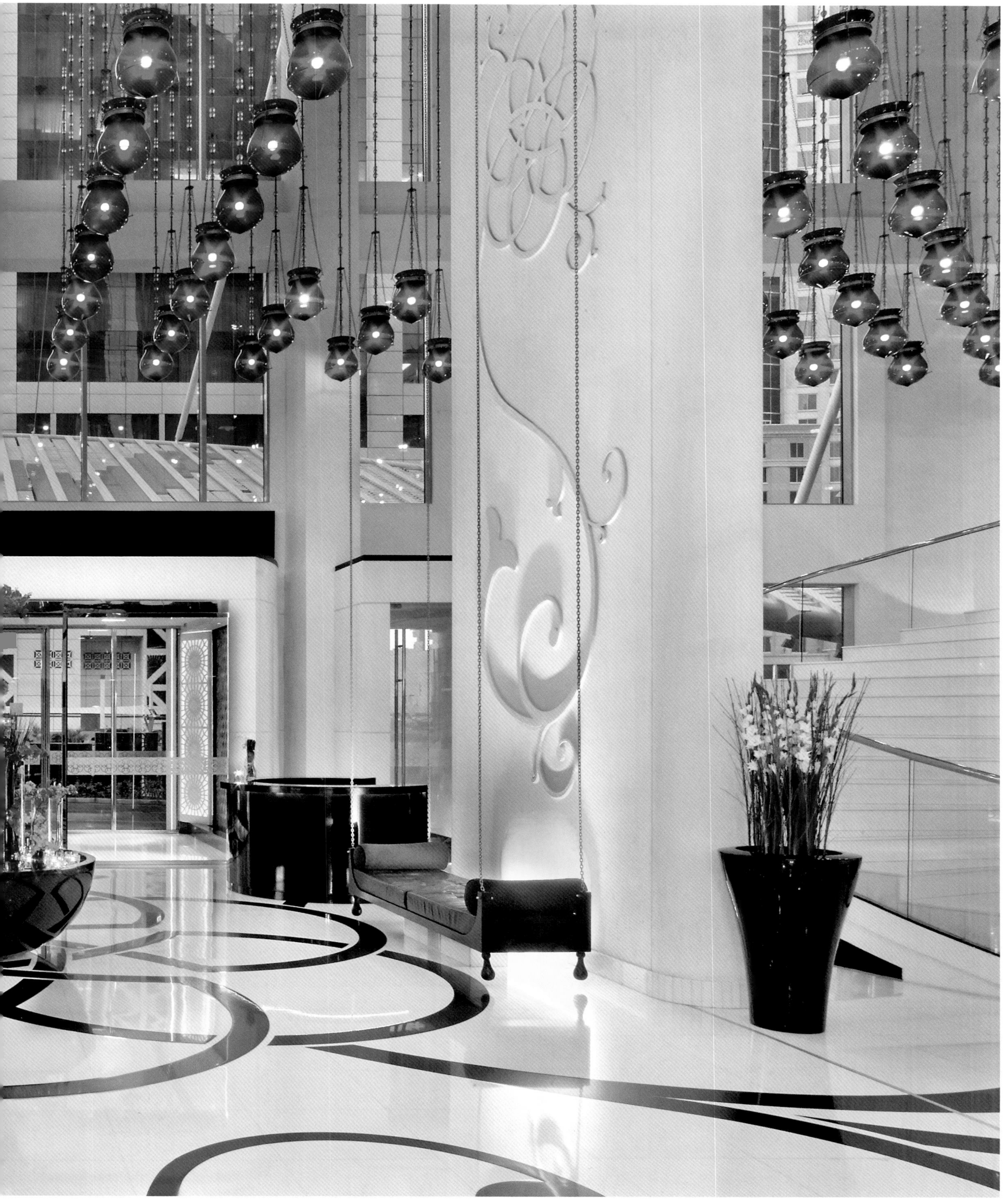

多哈君悦酒店 *Grand Hyatt Doha*

JW Marriott Absheron Baku 阿布歇隆区巴库JW万豪酒店

多哈瑞吉酒店 *The St. Regis Doha*

法拉克努马宫泰姬酒店 *Taj Falaknuma Palace, Hyderabad*

法拉克努马宫泰姬酒店
Taj Falaknuma Palace
Hyderabad

👑 **法拉克努马宫泰姬酒店** *Taj Falaknuma Palace, Hyderabad*

佛罗伦萨利马斯克雷别墅酒店 *Villa Le Maschere Resort, Florence*

佛山希尔顿酒店 *Hilton Foshan*

The Westin Fuzhou Minjiang 福州万达威斯汀酒店

哥印拜陀艾美酒店
Le Méridien Coimbatore

哥印拜陀艾美酒店 *Le Méridien Coimbatore*

广州 W 酒店
W Guangzhou

广州 W 酒店
W Guangzhou

古尔默尔格开柏喜马拉雅度假村及水疗中心
The Khyber Himalayan Resort & Spa, Gulmarg

121

广州花都合景喜来登度假酒店 *Sheraton Guangzhou Huadu Resort*

广州花都合景喜来登度假酒店 *Sheraton Guangzhou Huadu Resort*

海得拉巴威斯汀酒店 *The Westin Hyderabad Mindspace*

Banyan Tree Hangzhou 杭州西溪悦榕庄

海牙戴斯因德斯豪华精选酒店
Hotel Des Indes, a Luxury Collection Hotel, The Hague

129

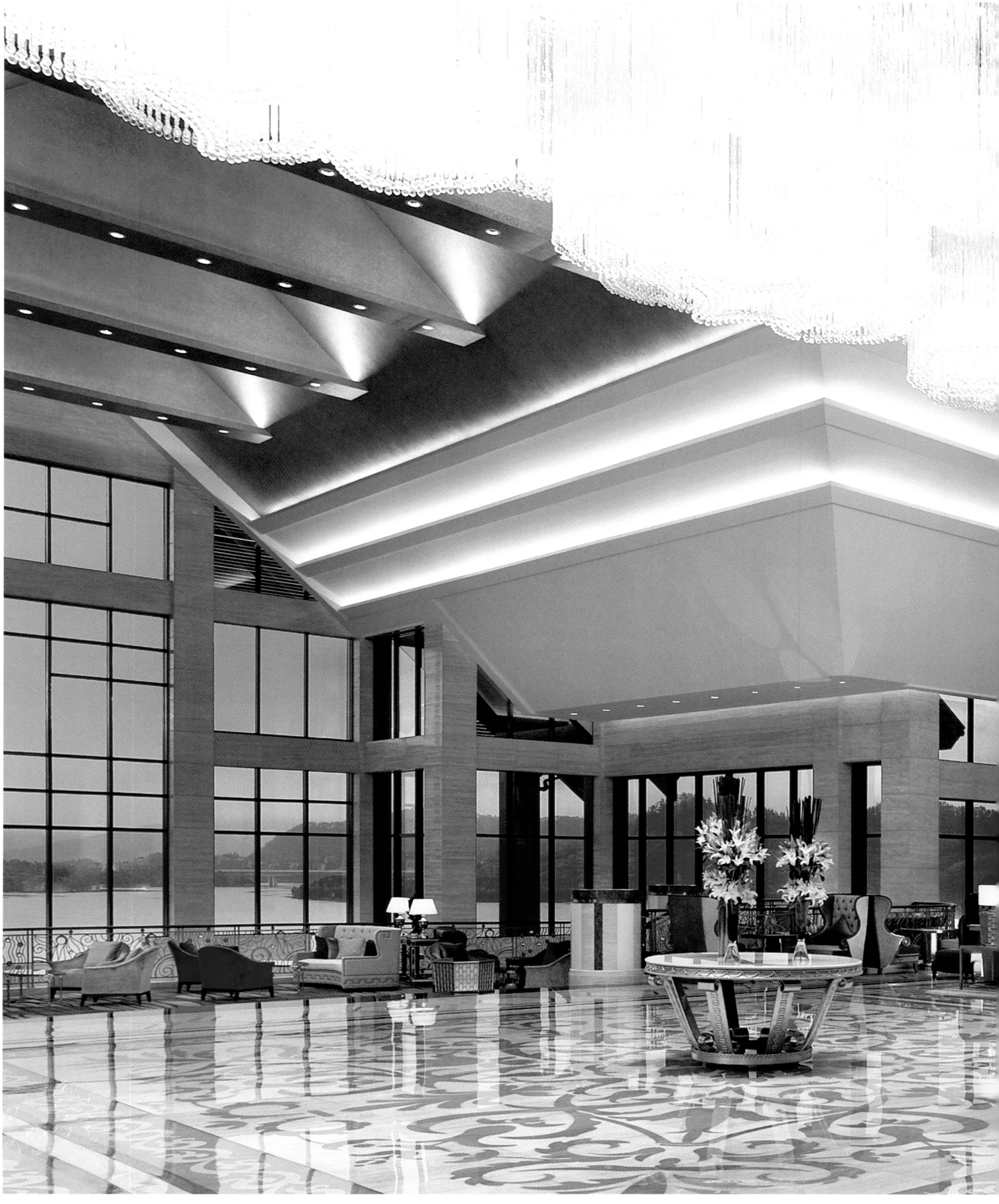

杭州千岛湖绿城喜来登度假酒店
Hilton Hangzhou Qiandao Lake Resort

杭州千岛湖滨江希尔顿度假酒店
Hilton Hangzhou Qiandao Lake Resort

杭州千岛湖滨江希尔顿度假酒店
Hilton Hangzhou Qiandao Lake Resort

杭州西溪喜来登度假酒店 *Sheraton Hangzhou Wetland Park Resort*

Sheraton Huzhou Hot Spring Resort 湖州喜来登温泉度假酒店

惠州金海湾喜来登度假酒店 *Sheraton Huizhou Beach Resort*

Sheraton Imperial Kuala Lumpur Hotel 吉隆坡喜来登帝王酒店

吉隆坡喜来登帝王酒店
Sheraton Imperial Kuala Lumpur Hotel

141

旧金山 W 酒店 *W San Francisco*

Wyndham Grand Plaza Royale Colorful Yunnan Kunming 昆明七彩云南温德姆至尊豪廷大酒店

旧金山瑞吉酒店 *The St. Regis San Francisco*

145

卡斯尔夏克华尔道夫酒店
Waldorf Astoria Jeddah-Qasr Al Sharq

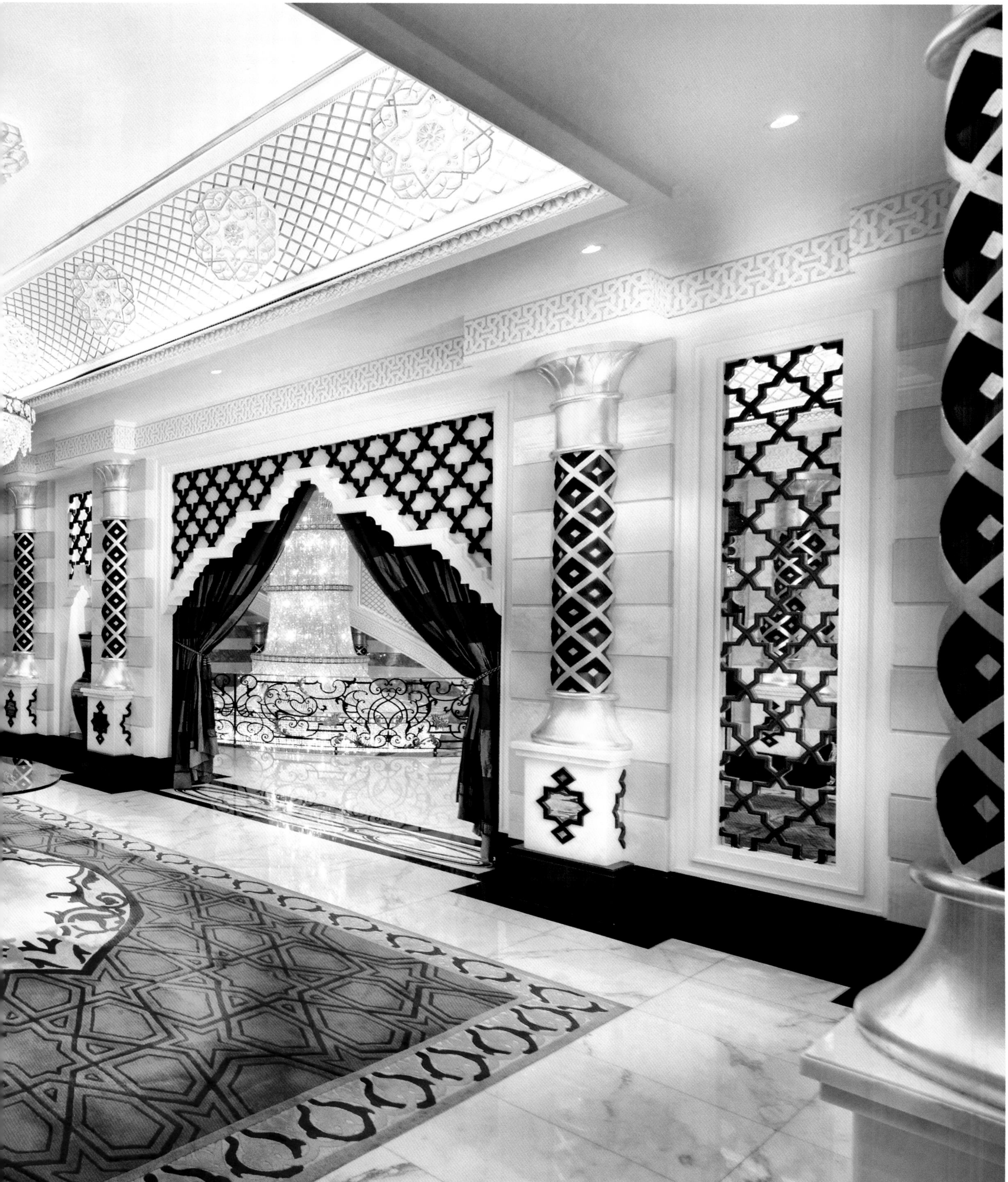

拉萨瑞吉度假酒店 *The St. Regis Lhasa Resort*

兰珂悦椿度假村
Angsana Lang Co

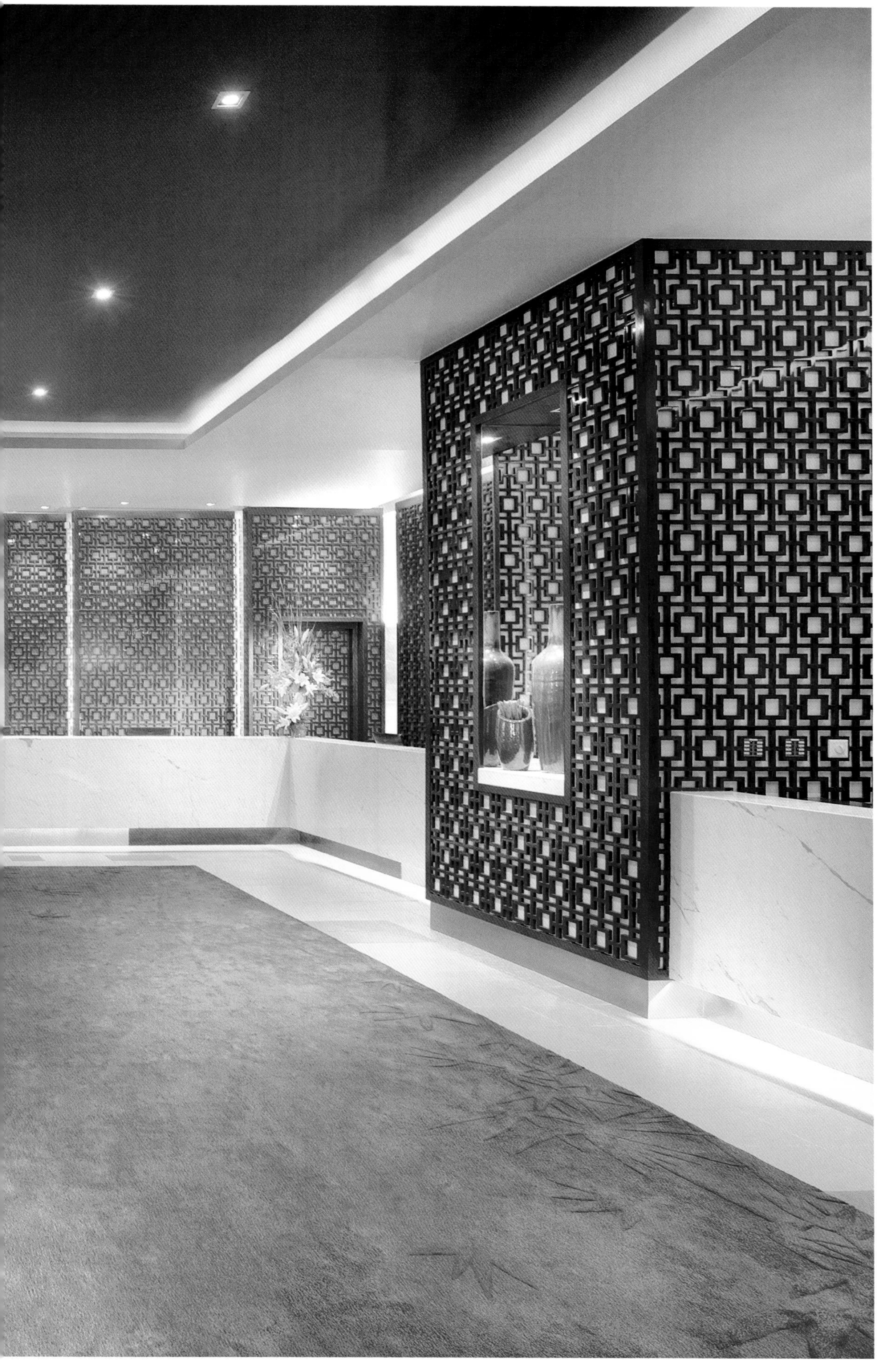

鹿谷瑞吉酒店
The St. Regis Deer Valley

鹿谷瑞吉酒店
The St. Regis Deer Valley

155

鹿谷瑞吉酒店
The St. Regis Deer Valley

157

马来西亚柔佛盛贸饭店 *Traders Hotel, Puteri Harbour, Johor, Malaysia*

The Westin Lima Hotel & Convention Center 利马威斯汀酒店及会议中心

罗得岛喜来登度假酒店 *Sheraton Rhodes Resort*

(Below) Hilton London Wembley （下图）伦敦希尔顿温布利酒店

马尔代夫瑞提拉岛 One&Only 度假村 *One&Only Reethi Rah, Maldives*

迈阿密南海滩 W 酒店 W South Beach

秘鲁唐波德尔英卡度假酒店
Tambo del Inka Resort & Spa, Valle Sagrado

秘鲁唐波德尔英卡度假酒店 *Tambo del Inka Resort & Spa, Valle Sagrado*

Le Méridien Ile Maurice 莫里斯岛艾美酒店

明尼阿波利斯威斯汀酒店
The Westin Minneapolis

纳泰攀牙麦拷梦水疗度假酒店 *Maikhao Dream Resort & Spa, Natai, Phang Nga*

（上图）南京威斯汀大酒店 *(Above) The Westin Nanjing*　（下图）米兰马尔蓬萨喜来登酒店及会议中心
(Below) Sheraton Milan Malpensa Airport Hotel & Conference Center

Grand Hyatt New York 纽约君悦酒店

宁波威斯汀酒店 *The Westin Ningbo*

纽约瑞吉酒店 *The St. Regis New York*

Sheraton Overland Park Hotel at the Convention Center 欧文兰德公园会议中心喜来登酒店

普林斯维尔瑞吉度假酒店
The St. Regis Princeville Resort

钦奈 ITC 大佐拉酒店 *ITC Grand Chola, a Luxury Collection Hotel, Chennai*

钦奈贝拉奇瑞威斯汀酒店 *The Westin Chennai Velachery*

Sheraton Qinhuangdao Beidaihe Hotel 秦皇岛北戴河华贸喜来登酒店

青岛胶州绿城喜来登酒店 *Sheraton Qingdao Jiaozhou Hotel*

(Above) Hyatt Regency Qingdao（上图）青岛鲁商凯悦酒店　　*(Below) Sheraton Shanghai Waigaoqiao Hotel*（下图）上海外高桥喜来登酒店

新加坡皮克林宾乐雅酒店 *Parkroyal on Pickering, Singapore*

(Above) Banyan Tree Tianjin Riverside（上图）天津海河悦榕庄　　*(Below)* Banyan Tree Ungasan（下图）乌干沙悦榕庄

清远狮子湖喜来登度假酒店 *Sheraton Qingyuan Lion Lake Resort*

Shangri-La Hotel, Qufu 曲阜香格里拉大酒店

曲阜香格里拉大酒店
Shangri-La Hotel, Qufu

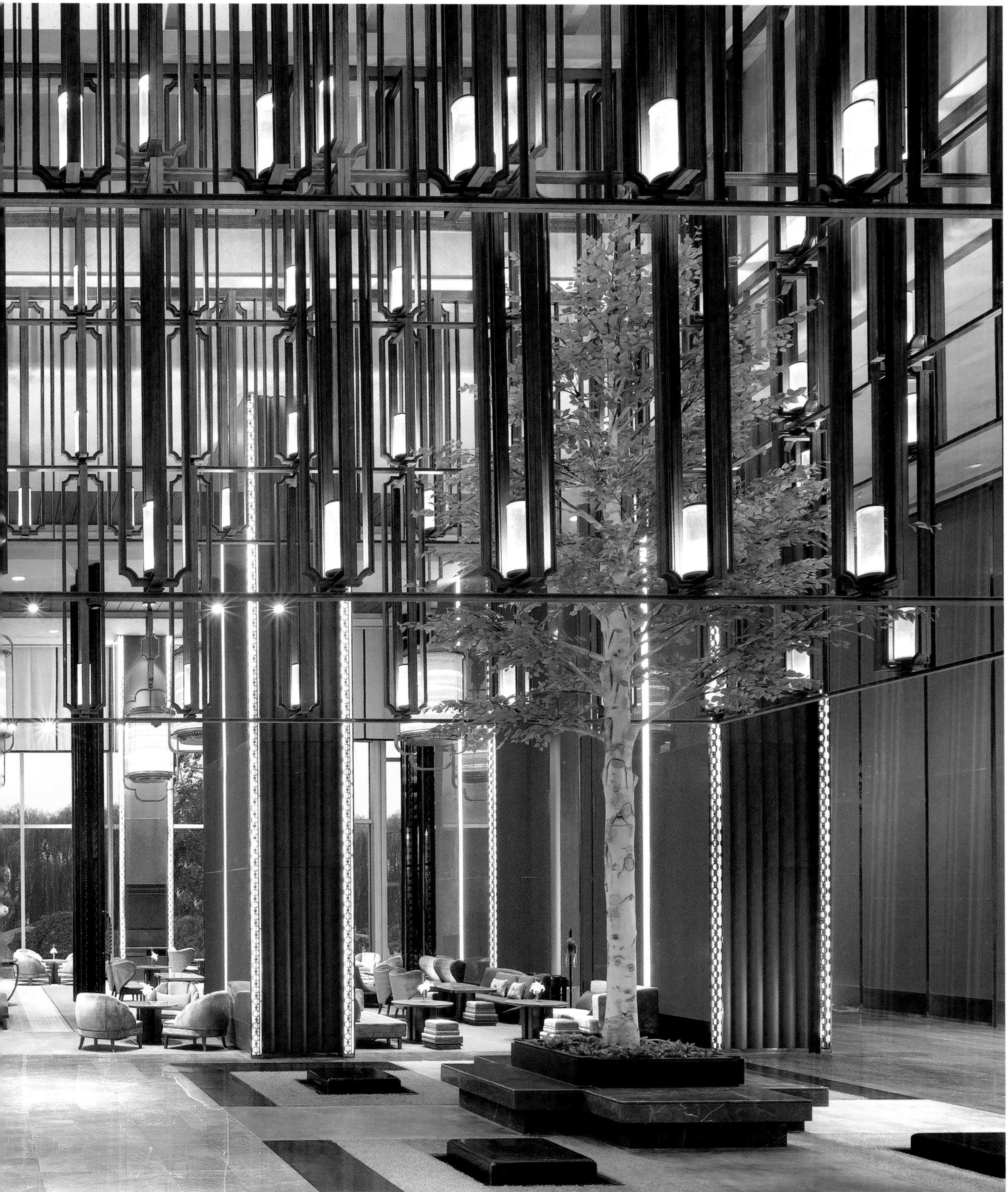

塞维利亚阿方索十三世豪华精选酒店
Hotel Alfonso XIII, a Luxury Collection Hotel, Seville

三亚海棠湾凯宾斯基酒店 *Kempinski Hotel Haitang Bay Sanya*

阿尔布费拉阿尔加维喜来登豪华精选酒店
Sheraton Algarve, a Luxury Collection Hotel,
Albufeira

三亚海棠湾喜来登度假酒店
Sheraton Sanya Haitang Bay Resort

三亚亚龙湾瑞吉度假酒店 *The St. Regis Sanya Yalong Bay Resort*

203

三亚御海棠豪华精选度假酒店 *The Royal Begonia, a Luxury Collection Resort, Sanya*

沙特阿拉伯王国麦加钟塔皇家费尔蒙酒店 *Makkah Clock Royal Tower, a Fairmont Hotel*

上海滴水湖皇冠假日酒店
Crowne Plaza Shanghai Harbour City

上海滴水湖皇冠假日酒店 *Crowne Plaza Shanghai Harbour City*

Jumeirah Himalayas Hotel Shanghai 上海卓美亚喜玛拉雅酒店

深圳东部华侨城瀑布酒店 Otique Aqua Hotel Shenzhen

(Above) Grand Hyatt Shenzhen （上图）深圳君悦酒店　　*(Below) Hilton Shijiazhuang* （下图）石家庄希尔顿酒店

深圳瑞吉酒店 *The St. Regis Shenzhen*

Sheraton Shenzhou Peninsula Resort 神州半岛喜来登度假酒店

阿布扎比盖斯尔阿萨拉沙漠度假村 *Qasr Al Sarab Desert Resort by Anantara, Abu Dhabi*

The Westin Sanya Haitang Bay Resort 三亚海棠湾民生威斯汀度假酒店

首尔 D 立方市喜来登酒店 Sheraton Seoul D Cube City Hotel

W Seoul - Walkerhill 首尔华克山庄 W 酒店

苏梅岛 W 酒店
W Retreat Koh Samui

索菲特曼谷特色酒店
Sofitel So Bangkok

台北 W 酒店 *W Taipei*

泰国清莱艾美度假酒店 *Le Méridien Chiang Rai Resort, Thailand*

天津瑞吉金融酒店
The St. Regis Tianjin

天津瑞吉金融街酒店
The St. Regis Tianjin

万绿湖东方国际酒店
Oriental International Hotel, Wanlv Lake, Heyuan

威尼斯丹尼利豪华精选酒店
Hotel Danieli, a Luxury Collection Hotel, Venice

威尼斯丹尼利豪华精选酒店
Hotel Danieli, a Luxury Collection Hotel, Venice

237

韦尔比耶 W 酒店
W Verbier

韦尔比耶 W 酒店 *W Verbier*

维也纳布里斯托尔豪华精选酒店 *Hotel Bristol, a Luxury Collection Hotel, Vienna*

香港愉景湾酒店 *Auberge Discovery Bay Hong Kong*

(Above) Anantara Xishuangbanna Resort & Spa
（上图）西双版纳安纳塔拉度假酒店

(Below) The Westin Wuhan Wuchang
（下图）武汉万达威斯汀酒店

西安赛瑞喜来登大酒店（城北）
Sheraton Xi'an North City Hotel

西双版纳避寒皇冠假日度假酒店
Crowne Plaza Resort Xishuangbanna

西双版纳避寒皇冠假日度假酒店
Crowne Plaza Resort Xishuangbanna

西双版纳避寒皇冠假日度假酒店
Crowne Plaza Resort Xishuangbanna

253

西双版纳避寒皇冠假日度假酒店 *Crowne Plaza Resort Xishuangbanna*

The Westin Sendai 仙台威斯汀酒店

厦门威斯汀酒店 *The Westin Xiamen*

香格里拉 Barr Al Jissah 度假酒店 *Shangri-La's Barr Al Jissah Resort & Spa, Sultanate of Oman*

259

香格里拉 Barr Al Jissah 度假酒店
*Shangri-La's Barr Al Jissah Resort & Spa,
Sultanate of Oman*

香格里拉莎利雅度假酒店 *Shangri-La's Rasa Ria Resort, Kota Kinabalu*

新德里 ITC 孔雀王朝豪华精选酒店 *ITC Maurya, a Luxury Collection Hotel, New Delhi*

新加坡圣淘沙湾 W 酒店 W Singapore - Sentosa Cove

新加坡万豪酒店
Singapore Marriott Hotel

269

271

新加坡威斯汀酒店 *The Westin Singapore*

The Westin Houston Downtown 休斯顿市中心威斯汀酒店

休斯顿市中心威斯汀酒店
The Westin Houston Downtown

休斯顿市中心威斯汀酒店
The Westin Houston Downtown

休斯顿市中心威斯汀酒店
The Westin Houston Downtown

雅典乔治国王豪华精选酒店
King George, a Luxury Collection Hotel, Athens

雅典乔治国王豪华精选酒店 *King George, a Luxury Collection Hotel, Athens*

283

雅加达苏丹皇宫广场酒店 *Keraton at The Plaza, Jakarta*

亚特兰大瑞吉酒店
The St. Regis Atlanta

亚特兰大瑞吉酒店
The St. Regis Atlanta

亚特兰大市中心 W 酒店
W Atlanta - Downtown

293

伊斯坦布尔艾美酒店 Le Méridien Istanbul Etiler

宜兴万达艾美酒店
Le Méridien Yixing

印度斋普尔费尔蒙酒店
Fairmont Jaipur, India

镇江万达喜来登酒店
Sheraton Zhenjiang Hotel

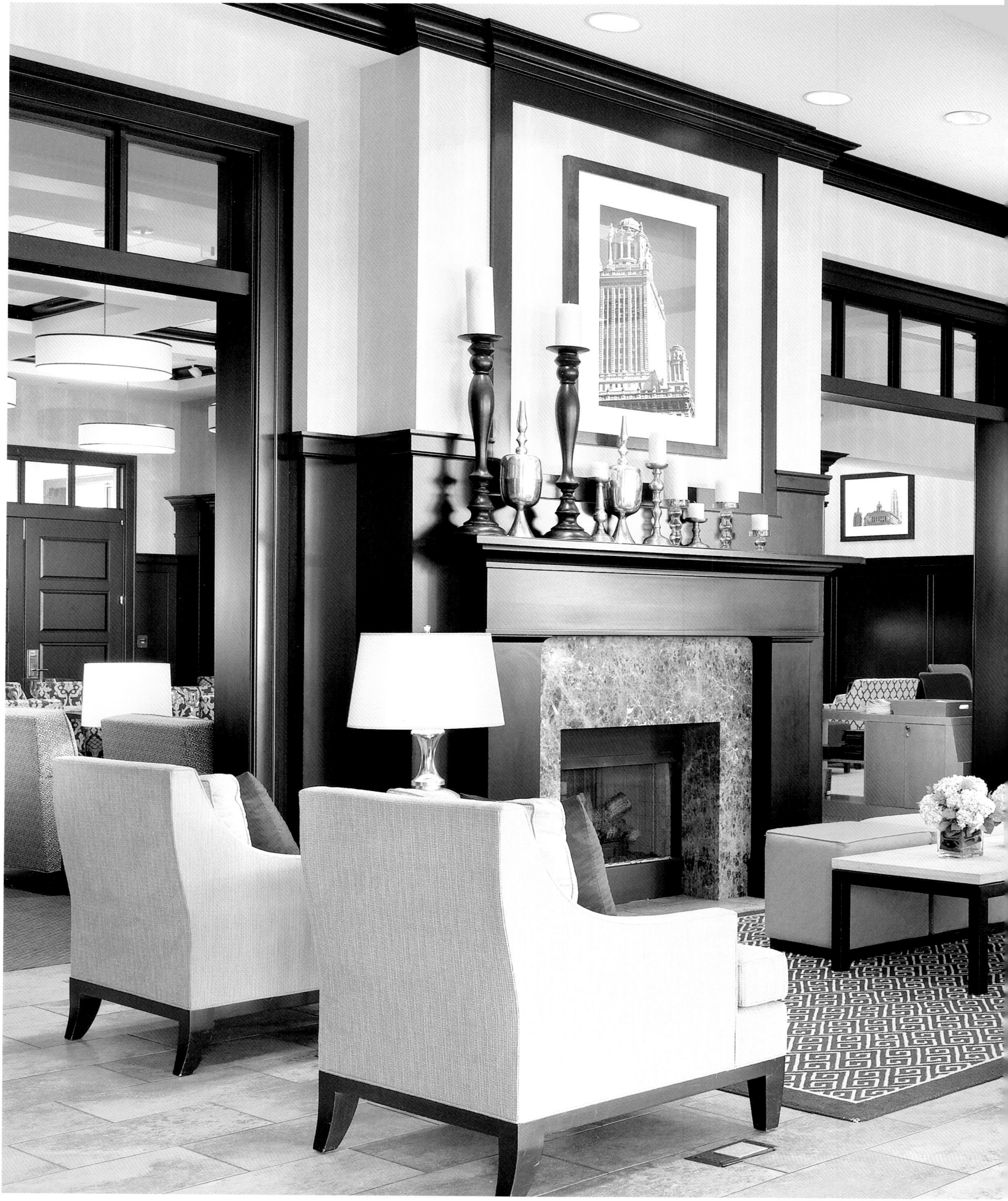

重庆北碚悦榕庄
Banyan Tree Chongqing Beibei

♛ 重庆北碚悦榕庄 *Banyan Tree Chongqing Beibei*

The Fullerton Bay Hotel Singapore 新加坡浮尔顿海湾酒店

卓美亚 Bilgah 海滩酒店
Jumeirah Bilgah Beach Hotel

卓美亚 Bilgah 海滩酒店 *Jumeirah Bilgah Beach Hotel*

卓美亚 Etihad Tower 酒店
Jumeirah at Etihad Towers

弗雷明汉喜来登会议中心酒店 *Sheraton Framingham Hotel & Conference Center*

Sheraton Maldives Full Moon Resort & Spa 马尔代夫满月岛喜来登度假酒店

迈阿密蒙德里安南部海滩酒店 *Mondrian South Beach Hotel Miami*

Sheraton Suites Calgary Eau Claire 卡尔加里奥克莱尔喜来登套房酒店

曼谷 W 酒店
W Bangkok

蒙特利尔 W 酒店 W Montréal

Le Méridien Chambers Minneapolis 明尼阿波利斯钱伯斯艾美酒店

柏林 *Axo Light in Rome and Bonaldo*
Axo Light in Rome and Bonaldo in Berlin

库斯科印加宫殿豪华精选酒店
Palacio del Inka, a Luxury Collection Hotel, Cusco

后记

酒店，对于有些人来说或许算得上是一个比较特殊而必须的"家"了，天南地北，来去匆匆，却也因这个临时的家而有莫大的安慰。行走于路上，感受的是缤纷的异域风情，最终也成就了不同文化的交融与渗透，而这种文化最好的承载体便是酒店，它丰富、包容。或许，在外地人看来，它总是有那么浓郁的地方特色；而于当地人来说，却因外地元素的注入而分外璀璨。行走不停，感受不止！

步入酒店大堂，所有的倦怠就完全消解于那一张柔软的沙发之中，忘却于流淌着的舒缓的钢琴声之中，如家的温馨，珍贵动人。堂皇气派的入户大堂，以殿堂级的奢装，超越名流贵胄的尊贵体验，或是精美壁画，或是巨大穹顶，或是雕花圆柱，极尽尊荣；内蕴十足的入户大堂，则以天然的视觉美学，尽享文人的诗意海洋，氤氲中国画，一曲流觞琴，便可雅享小树嫣然一两枝，正是微开半吐时的绝美意境，不论面纱存否，姿态可谓极为撩人……

李渔在其著作《笠翁一家言文集》里曾说："盖居室之前，贵精不贵丽，贵新奇大雅，不贵纤巧烂漫"，酒店大堂设计也就如此，而其氛围与意境的营造，则离不开对材料、色彩、灯光、陈设品等的运用。自20世纪50年代至今，现代酒店流行，新技术、新结构、新材料的飞速发展也为现代酒店建筑空间、环境塑造提供了无可比拟的自由度。酒店大堂内部设施的不断完善，内部空间布局的立体、多层次发展，现代技术与材料的利用等，都成就了大气尊荣、丰富多彩的现代酒店。

在酒店大堂设计之中，家具的陈设细分着大堂内部空间，它可划分出若干个不同的空间，既可有交流思想、情感，传递信息的公共之所，亦可有以供休憩的私享之地。家具的设计风格及其使用舒适度亦直接影响客人对整个大堂空间环境的感受，如沙发与地域手工艺品的结合，往往给人一种古朴的东方气息。壁饰与顶饰在酒店大堂设计中尤为重要，其墙面的材质与壁画不仅可以活跃空间气氛，增强空间文化氛围与艺术气息，而且可与普通墙面与顶面形成鲜明地对比，从而成为空间视觉焦点。酒店大堂是客人进入的第一个区域，其的高雅、明亮、宽敞、富丽的特质决定其一般以前暖色调为主，以使客人倍感温馨、舒畅。在设计酒店大堂时，无论是材料的运用，还是色彩的选择，皆服务于空间氛围的营造与整体观感的体验，于整个酒店形象与主题表现也尤为重要。

总的来说，一个成功的酒店大堂设计应以其精致的细节、舒适的氛围、专业的眼光，成就整个酒店的品位与辉煌。《酒店大堂设计》的出版，以期通过对众多经典酒店大堂设计地展示，开启设计师之灵感与感悟！

AFTERWORD

For some people, hotel is a special and necessary home! Course of this home, They have a great comfort on their short trip or long journey. Walking on the street, you will have a various exotic feeling. Also it contribute a blend and permeation of many kind of culture. The best carrier of the culture is hotel! Because it is abundant and comprehensive. Maybe in other people's view, the hotel has a strong distinctive feature. But the local think the hotel became brighter, because of the foreign elements addition. keep waling you will have lots of different feelings.

Walking in the hotel lobby, all the tiredness will disappear in the comfortable soft sofa, and forget in the eased piano. It fells like you were at home .this magnificent lobby is luxury like a palace, whatever the exquisite mural, the huge dome and the carved column in this lobby. You will have a more exalted experience than celebrity and nobility in there. Another different style lobby is specialize in naturally vision aesthetics, you will enjoy the poetic atmosphere there. The dense Chinese painting, the soothing music, and a tree with one or two branches, they all create a beautiful artistic conception.

Yu Li's Residential Comments, Habitable Room Design and Furnishings Arrangement recorded, before building a residence, we should put the key point on subtlety designation not gorgeous decoration, we should emphasize newness and elegance od design not small and exquisite. These words describe exactly what is lobby design. You can't create a great atmosphere and artistic conception without the use of material, color, light and furnishings. From 1950s, the fast development of modern hotel, new technology, new construction and new material provide an unparalleled degree of freedom for modern hotel space and environment. They all contribute a magnificent and colorful modern hotel, the continuous improvement of the hotel lobby interior, stereo and multi-level development of inner space layout, the use of modern technology and material and so on.

In the hotel lobby designation, The display of furniture subdivide the inner space in the lobby, it can divide the looby into many different style spaces. For example, a public place to exchange ideal, emotion and information; also a private place to sleep .the design style and comfort level of the furniture influence the customer's feeling of whole environment in the lobby. For instance the combination of sofa and local handiwork always gives people a kind of east plain feeling. The wall ornamentation and cresting is an important part of the lobby design. Material of the wall and mural is not only can create a lively atmosphere and enhance cultural and artistic atmosphere in the space, but also have a stark contrast to normal walls and top surface. So it is the visual focus of the whole space. The first area, the customer stepped in is lobby, it often use warm-toned to manifest it's elegance, brightness, capaciousness and magnificence. Also that can make customers feel warmer and more comfortable. Whatever the use of the material or the choice of color in the lobby design is in order to create atmosphere and experience overall impression, and it is also is a point to express image and theme of the hotel.

All in all, the delicate detail, comfortable atmosphere and professional designation of a successful hotel lobby design is to contribute the tates and glory of the hotel. *Hotel Lobby Design* comes out, hope to open your inspiration and gnosis through the expression of many classical top lobby designation.

图书在版编目（CIP）数据

酒店大堂设计 / DAM 工作室 主编．– 武汉：华中科技大学出版社，2014.5
ISBN 978-7-5680-0106-9

Ⅰ．①酒… Ⅱ．① D… Ⅲ．①饭店 – 室内装饰设计 Ⅳ．① TU247.4

中国版本图书馆 CIP 数据核字（2014）第 100140 号

酒店大堂设计　　　　　　　　　　　　　　　　　　　　DAM 工作室　主编

出版发行：华中科技大学出版社（中国·武汉）	
地　　址：武汉市武昌珞喻路1037号（邮编：430074）	
出 版 人：阮海洪	
责任编辑：段园园	责任监印：张贵君
责任校对：熊纯	装帧设计：筑美空间

印　　刷：中华商务联合印刷（广东）有限公司	
开　　本：965 mm × 1270 mm　1/16	
印　　张：20.5	
字　　数：164千字	
版　　次：2014年7月第1版 第1次印刷	
定　　价：338.00元（USD 67.99）	

投稿热线：（020）36218949　　　duanyy@hustp.com
本书若有印装质量问题，请向出版社营销中心调换
全国免费服务热线：400-6679-118 竭诚为您服务
版权所有　侵权必究